Exploring The BUILDING BLOCKS of SCIENCE

Book 2

TEACHER'S MANUAL

REBECCA W. KELLER, PhD

M000250783

Exploring the Building Blocks of Science Book 2 Teacher's Manual
ISBN 978-1-936114-36-8

Published by Gravitas Publications Inc.
www.realscience4kids.com
www.gravitaspublications.com

A Note From the Author

This curriculum is designed for elementary level students and provides an introduction to the scientific disciplines of chemistry, biology, physics, geology, and astronomy. *Exploring the Building Blocks of Science Book 2 Laboratory Notebook* accompanies the *Building Blocks of Science Book 2 Student Textbook.* Together, both provide students with basic science concepts needed for developing a solid framework for real science investigation. The *Laboratory Notebook* contains 44 experiments—two experiments for each chapter of the Student Textbook. These experiments allow students to further explore concepts presented in the *Student Textbook.* This teacher's manual will help you guide students through laboratory experiments designed to help students develop the skills needed for the first step in the scientific method — making good observations.

There are several sections in each chapter of the *Laboratory Notebook*. The section called *Think About It* provides questions to help students develop critical thinking skills and spark their imagination. The *Observe It* section helps students explore how to make good observations. In every chapter there is a *What Did You Discover?* section that gives the students an opportunity to summarize the observations they have made. A section called *Why?* provides a short explanation of what students may or may not have observed. And finally, in each chapter an additional experiment is presented in *Just For Fun*.

The experiments take up to 1 hour. The materials needed for each experiment are listed on the following pages and also at the beginning of each experiment.

Enjoy!

Rebecca W. Keller, PhD

Materials at a Glance

Experiment 1	Experiment 3	Experiment 4	Experiment 5	Experiment 6
magnifying glass butterfly or bug's wing (or substitute a leaf, flower, piece of wood, or rock) colored pencils microscope (or additional object to observe with a magnifying glass) **Experiment 2** salt, 15 ml (1 Tbsp.) water, 237 ml (1 cup) brick of modeling clay, 1 or 2 sugar	12 (or more) clear plastic cups measuring cup measuring spoons marking pen one head of red cabbage knife cooking pot, large food items: • distilled water, 1.25-1.75 liters (5-7 c.) • white grape juice, 60 ml (¼ cup) • milk, 60 ml (¼ cup) • lemon juice, 60 ml (¼ cup) • grapefruit juice, 60 ml (¼ cup) • mineral water, 60 ml (¼ cup) antacid tablets—3 extra-strength unflavored white Tums baking soda, 5 ml (1 tsp.) other substances (see *Just For Fun* section) **Optional** small plastic bag wooden mallet or other hard object for crushing antacids	18 or more clear plastic cups measuring cup measuring spoons marking pen leftover red cabbage juice from Experiment 3 or one head of red cabbage food items, approx 300 ml (1¼ c) each: • vinegar • lemon juice • mineral water • distilled water (if you need to make red cabbage juice, you will need 1.5 liters more) baking soda, 25 ml (5 tsp.) or more antacid tablets, 5 or more (try Tums plain, white, extra strength) substances of students' choice to mix together	the following food items: • marshmallows (2-3) • ripe banana • green banana • pretzels or salty crackers, several • raw potato • cooked potato • other food items blindfold	magnifying glass colored pencils

Experiment 7	Experiment 8	Experiment 9	Experiment 10	Experiment 11
microscope with a 10x or 20x objective lens (look online for sources such as Great Scopes) plastic microscope slides eye dropper pond water or protozoa kit[1] Protists (protozoa) can also be observed in hay water. To make hay water, cover a clump of dry hay with water and let it stand for several days at room temperature. Add water as needed.	(see Experiment 7) small piece of chocolate **Optional** baker's yeast Eosin Y stain[2] distilled water	6-8 sealable plastic bags waterproof disposable gloves piece of newspaper or plastic 2 pieces of fruit 2-3 pieces of bread (works best if bread does not have preservatives) marking pen water **Optional** colored pencils	clock or stopwatch	1 small glass marble 1 large glass marble

[1] As of this writing, the following materials are available from Home Science Tools, www.hometrainingtools.com: plastic microscope slides, MS-SLIDSPL or MS-SLPL144, Basic Protozoa Set, LD-PROBASC

[2] Eosin Y stain, CH-EOSIN

Experiment 12	Experiment 13	Experiment 14	Experiment 15	Experiment 16
stopwatch or clock an area to run in items for marking the beginning and ending of the running distance	4 plastic or Styrofoam cups with the mouth larger than the base 2 long poles (dowels work well or any two long sticks that are the same thickness from end to end) tape a cylinder, 10-13 cm long (4-5 inches) [such as a pencil, a dowel, a cylindrical block, or a cylindrical drinking glass that is not tapered; a paper towel tube may be used if it is filled with sand and the ends taped] chalk	plastic hammer regular metal hammer 3 pieces of banana 3 hardboiled eggs in the shell 3 raw potato halves 3 rocks of the same type and size (students can collect these) safety glasses **Optional** 8 pieces of paper marking pen	a toy, small music box, or toy car that can be taken apart a second similar item that can be taken apart screwdriver small hammer other tools as needed **Note:** The objects used in this experiment may not work again.	2 clear, tall glasses (drinking or parfait glasses) spoon (1 or more) 3-6 student-chosen food items for building a parfait model of Earth's layers (such as: graham crackers, peanut brittle, cookies, hot fudge, Jell-O, pudding, ice cream, cream cheese, cherry, nut, jelly bean. etc.) student-chosen inedible items that can be used to build a parfait model of Earth's layers (such as: rocks, mud, dirt, clay, dog or cat food, Legos, etc.) colored pencils

Experiment 17	Experiment 18	Experiment 19	Experiment 21	Experiment 22
colored pencils outdoor thermometer helium-filled balloon string	colored pencils clear night sky basketball or other large object(s) **Telescope materials*** empty cardboard paper towel tube 1-2 sheets of card stock or 1 manila file folder cut in half tape 2 lenses with different focal lengths from Home Science Tools: Item# OP-LEN4x15 and Item# OP-LEN4x50 http://www.hometrainingtools.com * Alternatively, you can look online for a telescope kit	colored pencils night sky daytime sky or textured surface **Optional** book or online information about constellations globe or basketball **Experiment 20** Styrofoam ball pick, awl, or other thin, sharp object to poke a hole through the center of the ball nylon string scissors 2 or more marbles of different sizes cups that are different sizes	flashlight with new batteries glow sticks in assorted colors may be found in places such as Walmart, toy stores, and online	10 small pieces of paper box for the paper pieces 2 beakers or jars: • one with 118 ml (½ cup) of vinegar • one with 118 ml (½ cup) of baking soda and water (5 ml [1 tsp] baking soda in 118 ml (½ cup) water) magnifying glass 2 balls of different weights (e.g., a glass marble and a metal marble, a plastic ball and a baseball) rock hammer or regular metal hammer safety glasses, 1 pair garden trowel or large metal spoon 10 pieces of paper 5 pens or pencils 4 friends or family members to help with the experiment scissors

Materials

Quantities Needed for All Experiments

Equipment	Foods	Foods (continued)
basketball or other large object(s) beakers or jars, 2 blindfold cooking pot, large eye dropper flashlight with new batteries glasses, safety, 1 pair hammer, plastic hammer, regular metal or rock hammer hammer, small knife magnifying glass, 1-2 measuring cup measuring spoons microscope with a 10x or 20x objective lens (look online for sources such as Great Scopes or Carolina Biological Supply) microscope slides, plastic [1] pick, awl, or other thin, sharp object to poke a hole through the center of a Styrofoam ball scissors screwdriver spoon (1 or more) stopwatch or clock thermometer, outdoor tools, misc. as needed trowel, garden or large metal spoon **Optional** mallet, wooden, or other hard object for crushing antacid tablets	antacid tablets—8 or more extra-strength unflavored white Tums baking soda, 35 ml (7 tsp.) or more banana, green, 1 banana, ripe, 2 bread, 2-3 pieces (bread without preservatives works best) cabbage, red, 1-2 heads cabbage juice, red, left over from Experiment 3 or one head of red cabbage to make new cabbage juice chocolate, small piece eggs, hardboiled in shell, 3 fruit, 2 pieces grape juice, white, 60 ml (¼ cup) grapefruit juice, 60 ml (¼ cup) lemon juice, 360 ml (1½ cup) marshmallows (2-3) milk, 60 ml (¼ cup) misc. food items misc. student-chosen food items for building a parfait model of Earth's layers (such as: graham crackers, peanut brittle, cookies, hot fudge, Jell-O, pudding, ice cream, cream cheese, cherry, nut, jelly bean. etc.) potato, cooked, 1 potato, raw, 3 pretzels or salty crackers, several salt, 15 ml (1 Tbsp.) sugar vinegar, 415 ml (1¾c) water, distilled, 1.5-3.5 liters (1.5-3.75 qt) or more water, mineral, 360 ml (1½ cup) water, tap	**Optional** baker's yeast

[1] As of this writing, the following materials are available from Home Science Tools, www.hometrainingtools.com: plastic microscope slides, MS-SLIDSPL or MS-SLPL144, Basic Protozoa Set, LD-PROBASC

Materials

Quantities Needed for All Experiments

Materials	Materials (continued)	Other
ball, Styrofoam balls of different weights (2), e.g., a glass marble and a metal marble, a plastic ball and a baseball balloon, helium-filled book or online information about constellations card stock, 1-2 sheets, or 1 manila file folder cut in half chalk clay, modeling, 1-2 bricks colored pencils cups, clear plastic, 30 or more cups, 4 plastic or Styrofoam, with the mouth larger than the base cups, several of different sizes cylinder, 10-13 cm long (4-5 inches) [such as a pencil, a dowel, a cylindrical block, or a cylindrical drinking glass that is not tapered; a paper towel tube may be used if it is filled with sand and the ends taped] glasses, 2 clear, tall drinking or parfait glasses gloves, waterproof disposable gloves, 2 pairs glow sticks in assorted colors—may be found in places such as Walmart, toy stores, and online items for marking the beginning and ending of a running distance items, misc.: student-chosen inedible items to use to build a parfait model of Earth's layers (such as: rocks, mud, dirt, clay, dog or cat food, Legos, etc.)	lenses (2) with different focal lengths Home Science Tools: Item# OP-LEN4x15 and Item# OP-LEN4x50 http://www.hometrainingtools.com (available as of this writing) * Alternatively, you can look online for a telescope kit marble, glass, 1 large marble, glass, 1 small marbles, 2 or more of different sizes newspaper or plastic, 2 pieces paper, 10 small pieces and box to put them in paper, 18 sheets or more paper towel tube, empty pen, marking pencils or pens, 5 pencils, colored plastic bags, sealable, 6-8 poles, 2 long (dowels work well or any two long sticks that are the same thickness from end to end) rocks, 3 of the same type and size (students can collect these) string, any string, nylon substances of students' choice to mix together tape toy, small music box, or toy car that can be taken apart and a second similar item that can be taken apart (they may not work again) **Optional** Eosin Y stain[2] globe (world) or basketball object (additional object to observe with a magnifying glass) plastic bag, small	area to run in butterfly or bug's wing (or substitute a leaf, flower, piece of wood, or rock) friends or family members (4) to help with experiment sky, clear night sky, daytime, or textured surface substances, other (see *Just For Fun* section, Experiment 3) water, pond or hay, or protozoa kit[1] Protists (protozoa) can be observed in hay water. To make hay water, cover a clump of dry hay with water and let it stand for several days at room temperature. Add water as needed.

[2] As of this writing, the following materials are available from Home Science Tools, www.hometrainingtools.com: Eosin Y stain, CH-EOSIN

How to Buy a Microscope

What to Look For

- A metal mechanical stage.
- A metal body painted with a resistant finish.
- DIN Achromatic Glass objective lenses at 4X, 10X, 40X (a 100X lens is optional but recommended).
- A focusable condenser (lens that focuses the light on the sample).
- Metal gears and screws with ball bearings for movable parts.
- Monocular (single tube) "wide field" ocular lens.
- Fluorescent lighting with an iris diaphragm.

Price Range

$50-$150: Not recommended: These microscopes do not have the best construction or parts and are often made of plastic. These microscopes will cause frustration, discouraging students.

$150-$350: A good quality standard student microscope can be found in this price range. We recommend Great Scopes for a solid student microscope with the best parts and optics in this price range. http://www.greatscopes.com

Above $350: There are many higher end microscopes that can be purchased, but for most students these are too much microscope for their needs. However, if you have a child who is really interested in microscopy, wants to enter the medical or scientific profession, or may become a serious hobbyist, a higher end microscope would be a valuable asset.

Objective lenses: Magnification/Resolution/Field of View/Focal Length

The objective lenses are the most important parts of the microscope. An objective lens not only magnifies the sample, but also determines the resolution. However, higher powered objective lenses with better resolution have a smaller field of view and a shorter focal length.

The resolution and working distance (focal length) of a lens is determined by its numerical aperture (NA). Following is a list of magnifications, numerical aperture, and working distance for some common achromatic objective lenses.

How to Buy a Microscope (Continued)

Magnification	Numerical Aperture	Working Distance (mm)
4X	0.10	30.00
10X	0.25	6.10
20X	0.40	2.10
40X	0.65	0.65
60X	0.80	0.30
100X (oil)	1.25	0.18

You can see as the magnification increases the numerical aperture increases (which means the resolution increases) and the working distance decreases.

Choosing the right lens for the right sample is part of the art of microscopy.

Most student projects can be achieved with a 40X objective, however a 100X objective lens can be added to make observing bacteria and small cell structures possible.

Below is a general chart showing the recommended objective lens to use for different types of samples.

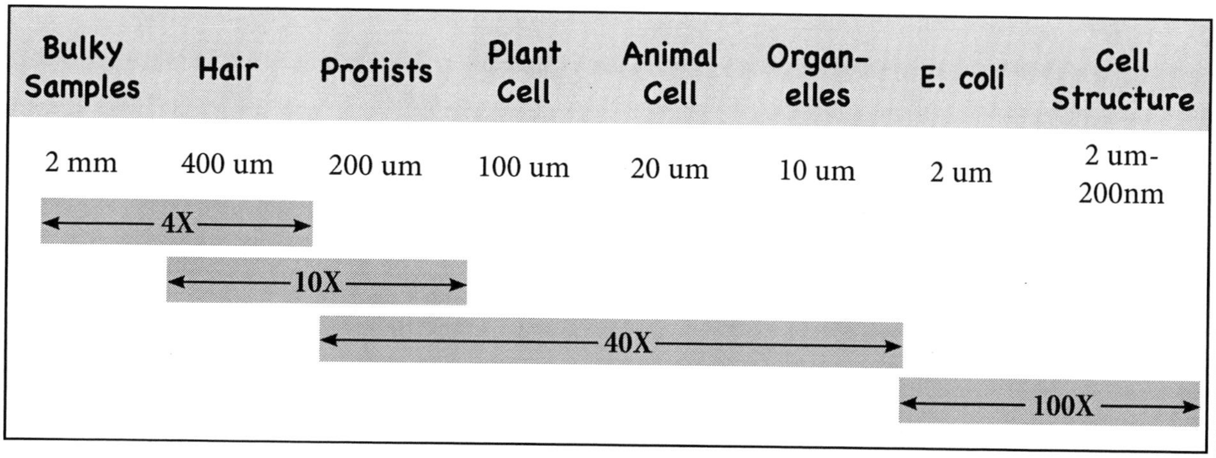

Contents

Experiment 1

Can You See?

Materials Needed

- magnifying glass
- butterfly or bug's wing (if not available, substitute a leaf, flower, piece of wood, or rock)
- colored pencils
- microscope (or additional object to observe with a magnifying glass)

Objectives

In this experiment students will explore how the use of basic tools can extend their senses and improve their observations.

The objectives of this lesson are for students to:

- Use suitable tools for making observations.
- Understand how tools help solve problems.

Experiment

I. Think About It

Read this section of the *Laboratory Notebook* with your students.

Have the students think about tools scientists use to explore the world around them. Discuss how the students' five senses give them information about the world around them and also how their senses are limiting.

Explore open inquiry with the following:

- *How far can you see with your eyes only?*
- *What is the smallest thing you can see with your eyes only?*
- *How many colors can you see with your eyes only?*
- *How many different smells can you smell?*
- *Do you think you can smell as well as a dog or cat? Why or why not?*
- *Do think a piece of paper is smooth or rough? How can you tell?*
- *Help the students think about how our senses give us some details about the world around us but that we cannot see everything there is to see with our eyes only. For instance, we can't detect very small changes in movement or resolve tiny features.*

There are no right answers to these questions.

Have the students think about a butterfly or bug's wing. If neither of these items can be found, use some natural item that has small details, such as a leaf, flower, piece of wood, or rock.

Help students think about a butterfly wing by asking questions such as the following:

- *What shape do you think a butterfly wing is?*
- *What color or colors do you think a butterfly wing is?*
- *Do you think it has a pattern?*
- *Do you think all butterfly wings look alike? Why or why not?*

In the space provided in the *Laboratory Notebook*, have the students draw their idea of what a butterfly wing (or substitute object) looks like *without looking at it first*. The name of the object can be written in the gray box.

There are no right answers. Let them use their imagination.

II. Observe It

Read this section of the *Laboratory Notebook* with your students.

Have your students observe the details of a butterfly wing, bug's wing, or substitute object without the use of a magnifying lens. In the space provided in the *Laboratory Notebook*, have students record what they observe.

Next, have your students observe the same item with the magnifying lens. Encourage them to look at details they could not see with their eyes only. Have them record what they see.

III. What Did You Discover?

Read this section of the *Laboratory Notebook* with your students.

❶-❹ Have the students answer the questions. These can be answered orally or in writing. Again, there are no right answers and their answers will depend on what they actually observed.

IV. Why?

Read this section of the *Laboratory Notebook* with your students.

Discuss any questions that might come up.

V. Just For Fun

Have the students observe the same item with a microscope. (If a microscope isn't available, have the students find another object to look at first with their eyes and then using the magnifying glass. Guide them in finding an object that has details that can be seen.)

Have the students record their observations in the spaces provided.

Experiment 2

The Clay Crucible

Materials Needed

- salt, 15 ml (1 Tbsp.)
- water, 237 ml (1 cup)
- brick of modeling clay (1 or 2)
- sugar

Objectives

In this experiment, students will explore how to make and use a basic chemistry tool.

The objectives of this lesson are for students to:

- Create a suitable tool to use in experiments.
- Understand how tools help solve problems.

Experiment

Introduction

Read this section of the *Laboratory Notebook* with your students.

Discuss any questions students may have.

I. Think About It

Read this section of the *Laboratory Notebook* with your students.

Have the students think about how chemistry tools help chemists do specific experiments. Discuss how beakers help chemists measure the volume of liquids and how a balance or scale helps chemists measure the weight of solids.

Explore open inquiry with questions such as the following:

- *Can you hold liquid water in your hand? Why or why not?*
- *Can you hold pebbles in your hand? Why or why not?*
- *Can you use a scale to measure liquid water by pouring water on the scale? Why or why not?*
- *Do you think you could use a beaker to measure large pebbles? Why or why not?*

Have the students think about how these two different tools (beakers and scales) are used by chemists to measure different types of chemicals—liquid chemicals and solid or dry chemicals.

Have the students think of some ways that might be used to separate the salt out of saltwater, and then have them record their ideas in the box provided. Let them use their imagination. There are no right answers to this section.

II. Observe It

Read this section of the *Laboratory Notebook* with your students.

A crucible is a dish-shaped or cup-shaped tool used in chemistry labs. Have your students create a small crucible from the clay. Guide them to notice that they can create a shallow crucible or a deep one. Discuss with them how a shallow crucible will allow the water to evaporate more quickly and a deep, narrow crucible will slow the evaporation down. Have them create a suitable crucible.

Have the students mix the saltwater, making sure the salt is completely dissolved. Have them pour the saltwater into the crucible and then observe what happens over the course of several hours or days. In hot, dry climates, the water will evaporate more quickly, and in cold, humid environments, the water will evaporate more slowly. Have them record what they observe, noting the date and the time of the observation. They may make their observations on the same day or on different days.

III. What Did You Discover?

Read this section of the *Laboratory Notebook* with your students.

❶-❹ Have the students answer the questions. These can be answered orally or in writing. There are no right answers and their answers will depend on what they actually observed.

IV. Why?

Read this section of the *Laboratory Notebook* with your students.

Discuss any questions that might come up.

V. Just For Fun

Have the students repeat the experiment with a sugar/water mixture or a salt/sugar/water mixture. If there is enough clay left over, they might like to make another crucible, trying both mixtures and comparing the results. By adding the sugar (and salt, if used) a little at a time, they can experiment to see how much can be added before it will no longer dissolve.

They are asked to first record what they think will happen and then what actually happens during the experiment.

Experiment 3

Sour or Not Sour?

Materials Needed

- 12 (or more) clear plastic cups
- measuring cup
- measuring spoons
- marking pen
- one head of red cabbage
- knife
- cooking pot, large
- the following food items:
  distilled water, 1.25–1.75 liters
  (5–7 cups)
  white grape juice, 60 ml (1/4 cup)
  milk, 60 ml (1/4 cup)
  lemon juice, 60 ml (1/4 cup)
  grapefruit juice, 60 ml (1/4 cup)
  mineral water, 60 ml (1/4 cup)
  antacid tablets—3 extra-strength
  unflavored white Tums
  baking soda, 5 ml (1 teaspoon)
- other substances to test
  (see *Just For Fun* section)

Optional

- small plastic bag
- wooden mallet or other hard object
  for crushing antacid tablets

CHEMISTRY

Objectives

In this experiment students will begin to explore the properties of acids and bases.

The objectives of this lesson are:

- To have students observe that acids and bases have different properties.
- To introduce the concept of indicators — in this experiment, red cabbage juice is used as an acid-base indicator to determine whether liquids are acids or bases.

Experiment

This experiment requires that students taste both acids and bases. It is relatively easy to find foods that are acidic but much more difficult to find foods that are basic. The only two safe products that we could find that are basic are baking soda and antacids. Most household cleaning products are basic, but these are not listed since they are not safe to taste.

Setup

NOTE: Do not use tap water for this experiment. Use only distilled water or you will not get the correct results.

To do 1 hour before the experiment

Chop or shred the head of red cabbage, and boil it in 1-1.5 liters (4-6 cups) of distilled water for 15 minutes. Remove the cabbage and allow the liquid to cool to room temperature.

Prepare liquids to be tested.

Dissolve the antacid tablets in distilled water. Add three extra-strength unflavored white Tums tablets to 60 milliliters (1/4 cup) of distilled water. Crushing the tablets may help in obtaining the color change of the cabbage juice indicator. To crush them, the tablets can be put in a plastic bag and hit with a hard object, such as a wooden mallet. If this mixture does not change the color, try adding more tablets. Other brands of antacids may or may not work.

To make the baking soda water, add 5 milliliters (1 teaspoon) of baking soda to 60 milliliters (1/4 cup) of distilled water.

Pour 60 milliliters (1/4 cup) of each liquid into a separate clear plastic cup, and using a marking pen, label each cup with the name of its contents.

I. Think About It

Read the text with your students.

Have the students make predictions about which liquids will taste sour and which will not taste sour. Help them mark their predictions in the proper column of the table in their *Laboratory Notebook*. Their answers may vary.

II. Observe It

Read this section of the *Laboratory Notebook* with your students.

❶ Have the students tear out the *Laboratory Notebook* pages that are labeled "SOUR" and "NOT SOUR" and place them on a table.

Have the students taste each liquid and indicate on the chart in the *Laboratory Notebook* whether it is sour or not sour. Help them try to distinguish between "sour" and "bitter." The mineral water and the baking soda water will taste "bad" but not sour. They are bitter or salty. The antacid water will taste sweet. Also, white grape juice may be sweet and not necessarily sour. Let the students decide whether they think it is sour or not sour. After recording each answer, have them place the cup of liquid on either the SOUR or NOT SOUR page, according to the taste.

Their answers may look as follows (answers may vary).

Liquid	Sour	Not Sour
white grape juice		X
milk		X
lemon juice	X	
grapefruit juice	X	
mineral water		X
antacid		X
distilled water		X
baking soda water		X

❷ Pour into a measuring cup the red cabbage juice that you made earlier and have the students observe the color of the cabbage juice. Then have them select one of the cups of liquid they tasted and observe the color of that liquid. Next, have them add 60 milliliters (1/4 cup) of red cabbage juice to the liquid.

❸ Ask them whether the color changes or stays the same. What they are looking for is ***the color change of the cabbage juice***. Its natural color is a deep red-purple. It will change to pink, green, or light purple when mixed with the other liquids.

Have the students return the cup to the SOUR or NOT SOUR page they took it from, and have them record their results.

❹ Have the students repeat Steps ❷-❸ for each of the liquids they tasted. Expected results are shown in the following chart:

Liquid	Color change? (yes or no)	What is the color?
white grape juice	*yes*	*pink*
milk	*no*	*purple*
lemon juice	*yes*	*pink*
grapefruit juice	*yes*	*pink*
mineral water	*yes/no*	*light purple*
antacid	*yes*	*green*
distilled water	*no*	*purple*
baking soda water	*yes*	*green*

III. What Did You Discover?

Have the students look at the cups that are on the SOUR and NOT SOUR pages. Ask them to observe the colors of the liquids and whether they see similarities or differences between those that are on the same page.

Help the students answer the questions in this section. Example answers follow.

(Answers may vary.)

1. Which liquids were sour? *lemon juice and grapefruit juice*
2. Which liquids were not sour? *milk, distilled water, mineral water*
3. When you added the cabbage juice to the "sour" liquids, what color did the cabbage juice become? *pink*
4. When you added the cabbage juice to the "not sour" liquids, what color did the cabbage juice become? *green or purple*
5. Why do you think the "sour" liquids and "not sour" liquids turned the cabbage juice different colors? *They have different types of molecules.*
6. If you added cabbage juice to a drink and it turned pink, do you think that drink would taste sour? *yes*

IV. Why?

Read the text with your students.

Discuss this section with the students. Have them think about why some of the liquids turned the red cabbage juice pink and some turned it green. Explain to them that the liquids that turned the cabbage juice pink are called acids, and the liquids that turned the cabbage juice green are called bases.

Explain that red cabbage juice is an indicator, which is anything that points out something to us. For example, a gas gauge in a car could be called an indicator—it tells the level of gas in the tank. The thermostat in a house could be called an indicator—it tells the temperature of the room.

In chemistry the term *indicator* refers to a chemical that tells you something about other chemicals. Red cabbage juice is an acid-base indicator, telling you whether the liquid is acidic or basic.

Explain that red cabbage juice will always turn pink in acids and will always turn green in bases unless there is something wrong with the indicator. Some liquids, such as milk and water, do not turn the indicator another color. Explain that these liquids are called *neutral*, and they are neither acids nor bases.

V. Just For Fun

Help the students find some other liquids to test with the red cabbage juice indicator. The students are **NOT TO TASTE** these liquids, so they can select some things like household cleaners that are not edible. They can also mix a powdered substance into distilled water and test the mixture.

Have the students decide whether the liquid is an acid, a base, or neutral, and help them record their observations.

Experiment 4

Pink and Green Together

Materials Needed

- 18 or more clear plastic cups
- measuring cup
- measuring spoons
- marking pen
- leftover red cabbage juice from Experiment 3 or one head of red cabbage
- the following food items, approx. 300 ml (1 1/4 cups) each:
  vinegar
  lemon juice
  mineral water
  distilled water (if you need to make red cabbage juice, you will need 1.5 liters more)
- baking soda, 25 ml (5 tsp.) or more
- antacid tablets, 5 or more (try Tums plain, white, extra strength)
- substances of students' choice to mix together

Objectives

In the last experiment students added red cabbage juice to several liquids to determine which were acids and which were bases. In this experiment students will continue their exploration of acids and bases.

The objectives of this lesson are for students to:

- Explore what happens when an acid and a base are mixed together.
- See that mixing an acid and a base can result in a neutral mixture, one that is neither an acid nor a base.

Experiment

If you have refrigerated red cabbage juice left from Experiment 3, use that. Otherwise follow the directions below.

If you do not have red cabbage juice from Experiment 3, do the following 1 hour before:

Chop or shred one head of red cabbage and boil it in approximately 1.5 liters (6 cups) of distilled water for 15 minutes. Remove the cabbage and allow the liquid to cool to room temperature.

NOTE: Do not use tap water. Use only distilled water or you will not get the correct results.

Setup

Put 60 ml (1/4 cup) of each of the following liquids into separate plastic cups and label the cups with a marking pen:

- vinegar
- lemon juice
- mineral water
- distilled water

Take two more plastic cups and put 60 ml (1/4 cup) of distilled water in each. Add 5 ml (1 tsp.) of baking soda to one cup and an antacid tablet to the other. (You may want to break or crush the tablet to help it dissolve faster.)

Alternatively, you can mix enough baking soda water and antacid water for the entire experiment. Use 300 ml (1 1/4 cups) distilled water to 25 ml (5 tsp.) baking soda and the same amount of distilled water with 5 or more antacid tablets. Then put 60 ml (1/4 cup) of each solution in a cup.

Label the cups.

I. Think About It

Read this section of the *Laboratory Notebook* with your students.

❶-❺ Have the students think about and answer the questions in this section of the *Laboratory Workbook*. Their answers will vary.

II. Observe It

Read this section of the *Laboratory Notebook* with your students.

❶ Place all of the cups on the table and have the students add 60 ml (1/4 cup) of cabbage juice to each cup. Have them observe the color of the liquid in each cup, and then have them record their results in the chart.

They should get the following:

Liquid	Pink	Green	Purple
distilled water			*X*
mineral water			*X*
lemon juice	*X*		
vinegar	*X*		
baking soda water		*X*	
antacid water		*X*	

❷ Have the students pour a green liquid into a cup containing a pink liquid and then a pink liquid into a cup containing a green liquid. This way they can observe that the result will be the same if an acid is added to a base or a base is added to an acid. Have them try all the combinations of pink and green liquids. Help the students fill more cups as needed.

Have them record their observations. Not all the empty squares will be filled in during this part of the experiment.

	antacid water	lemon juice	vinegar	mineral water	distilled water	baking soda water
antacid water	X					
lemon juice	X	X				
vinegar	X	X	X			
mineral water	X	X	X	X		
dis-tilled water	X	X	X	X	X	
baking soda water	X	X	X	X	X	X

❸ Next, have the students mix together the remaining liquids listed on the chart. Help them record any color changes that occur. For example, when lemon juice (pink) is added to mineral water (purple), the mineral water will turn pink. When mineral water (purple) is added to baking soda water (green), the color may change only slightly.

Encourage them to keep pouring the liquids back and forth to see what happens when mixtures are added to other mixtures. In the end, all of the liquids should turn purple. If some liquids are still green or pink, have the students pour them back and forth until every cup contains purple liquid.

Have them record anything they observe that they find interesting.

III. What Did You Discover?

Read this section of the *Laboratory Notebook* with your students.

❶-❹ Help the students answer the questions in this section. They should have seen some of the pink liquids turn green when green liquid was added and some of the green liquids turn pink when pink liquid was added.

IV. Why?

Read this section of the *Laboratory Notebook* with your students.

Have the students look at the chart they made and discuss the results with them. Explain that when they poured the liquids back and forth, the colors changed because the acids and bases were *reacting* with each other. Remind the students that in Chapter 3 they learned that a chemical reaction can be *indicated* by a color change. In this experiment the red cabbage juice indicator changed color as the acids and bases reacted with each other.

At the end of the experiment all of the liquids turn purple. Explain to the students that the acids and bases react with each other and cancel each other out, or *neutralize* each other. In the end there are no acids or bases left, only neutral liquids.

V. Just For Fun

Have the students look for some different liquids they can mix together and test with red cabbage juice indicator. Have them save these mixtures and then mix the mixtures together and see what happens. Do they change color? Have them observe whether they have an acid, a base, or a neutral mixture. They are not to taste these mixtures.

Experiment 5

Salty or Sweet?

Materials Needed

- the following food items:
  marshmallows (2-3)
  ripe banana
  green banana
  pretzels or salty crackers,
    several
  raw potato
  cooked potato
  other food items
- blindfold

Objectives

In this experiment students will explore the concept that foods flavored by different molecules taste different.

The objectives of this lesson are to introduce students to the concepts that:

- Different molecules create different flavors in foods.
- The tongue has taste buds that sense different flavors.
- Long chains of carbohydrate molecules must be broken apart in order for the tongue to be able to taste the sugar molecules.

Experiment

To do 1 hour before:

Boil a raw potato until soft, then mash it and let it cool.

I. Think About It

Read this section of the *Laboratory Notebook* with your students.

Guide open inquiry with questions such as the following. There are no right answers to these questions.

- *Do you think you could taste anything if you did not have a tongue? Why or why not?*
- *Do you think you would enjoy eating food if everything tasted the same? Why or why not?*
- *How many different tastes do you think your tongue can detect?*
- *Do you think it makes a difference in the flavor of foods like potatoes and carrots if they are cooked or raw? Why or why not?*
- *If you chopped up raw vegetables, do you think they would taste different from when they are whole? Why or why not?*
- *What other indicators does your body have? (Guide the students to think of their five senses and go from there.)*

❶-❻ Ask the students to think about the questions in this section of the *Laboratory Notebook*. Help them record their answers. There are no right answers, and their answers may vary from those you would expect.

II. Observe It

Read this section of the *Laboratory Notebook* with your students.

Have the students tear out the pages labeled **SALTY**, **SWEET**, and **NEITHER** and spread them out on a table. Provide the food items to be tasted.

Have the students guess, *without tasting*, which foods will be salty, which will be sweet, and which will be neither. Have them place the foods on the corresponding pages.

Now take a blindfold and cover the students' eyes. Hand them one of the items from one of the pages, and ask them to guess if it is a sweet item, a salty item, or neither sweet nor salty. If they guess correctly, place the item back on the labeled paper. If their guess was incorrect, place the item off to the side.

When they finish tasting the items, remove the blindfold and have them see how many items they guessed correctly.

III. What Did You Discover?

Read this section of the *Laboratory Notebook* with your students.

❶-❻ Discuss the questions in this section. Help the students record their answers. Ask them how many foods they guessed correctly and how many they didn't guess correctly.

IV. Why?

Read this section of the *Laboratory Notebook* with your students.

Explain that the different flavors we taste in foods come from different molecules. The tongue is designed to detect these different molecules, causing the experience of different flavors. The taste buds on the tongue can tell salt molecules from sugar molecules.

Have a discussion about the fact that foods such as raw potatoes and green bananas contain long chains of sugar molecules called carbohydrates. Because the sugar molecules in carbohydrates are hooked together in long chains, taste buds cannot detect them. This is why raw potatoes and green bananas do not taste very sweet. Explain that when potatoes are cooked and bananas ripen, they become sweeter than when they are uncooked or not ripe. Cooking and ripening break apart carbohydrates (the long chains of sugar molecules) and then taste buds can detect the sugar.

V. Just For Fun

Have the students see if they get the same results if they repeat the experiment with someone else tasting the same foods. They might also like to find additional foods for comparison taste testing. Have them record their observations in the space provided.

Experiment 6

Taking Notes

Materials Needed

- magnifying glass
- colored pencils

Objectives

In this experiment, students will explore how to use their five senses and basic tools to make good observations.

The objectives of this lesson are for students to:

- Observe both natural and man-made objects.
- Use a simple tool to make better observations.

Experiment

I. Think About It

Read this section of the *Laboratory Notebook* with your students.

Have the students think about how making good observations is important for studying biology. Have them explore how the natural world is both predictable and unpredictable and how plants and animals can react to the environment in both expected and unexpected ways (e.g., a cat may usually chase a mouse, but a cat might also just watch the mouse move and choose not to chase it).

Explore open inquiry with the following:

- *What does an ant look like? What color, shape, and size is it?*
- *What do you think is inside a tree? What does it look like? What color is the inside? Do you think there are several colors or just one?*
- *What happens when a cat jumps on a wall? How does it move its back legs? Does it use its front legs?*
- *What is the shape of a bird wing? Are all bird wings the same shape and size? Why or why not?*

Have the students think about these questions. Then have them come up with five of their own biology questions. There are five boxes provided in the *Laboratory Notebook* for their questions. Have the students write each question in the gray part of a box and then in the area below the question, write or draw their ideas about possible answers. There are no right answers to their questions. Encourage the students to use their imagination.

II. Observe It

Read this section of the *Laboratory Notebook* with your students.

Have the students take their *Laboratory Notebook*, pen or pencil, and a magnifying glass and go outside. They are not to use any type of electronic device to record their observations. They are also not to use a cellphone during this time. Using the five senses in doing science is a skill that needs to be developed before adding the use of technological aids. Have the students sit quietly for 10 minutes or so and just observe the smells, colors, sounds, and textures of the space surrounding them. Encourage them to use the magnifying glass to view any details they find interesting.

In the boxes provided in the *Laboratory Notebook*, have the students record what they see, hear, smell, and touch in the space around them. If they are sitting on the ground with dirt or grass, have them record what kind of ground they're on. If they are sitting on a bench or chair, have them record this. Have them note which of the items they observe are natural and which items are man-made.

III. What Did You Discover?

Read this section of the *Laboratory Notebook* with your students.

❶-❼ Have the students refer to their notes in the *Observe It* section and answer the questions. These can be answered orally or in writing. Again, there are no right answers and their answers will depend on what they actually observed.

IV. Why?

Read this section of the *Laboratory Notebook* with your students.

Discuss any questions that might come up.

V. Just For Fun

Read this section of the *Laboratory Notebook* with your students.

Have the students repeat this exploration in a place they have never visited. Take them to a new park or outdoor space. Or you can take them to a museum, library, ice cream parlor, or other location. Have them record what they see, smell, and hear.

Experiment 7

Little Creatures Move

Materials Needed

- microscope with a 10x or 20x objective lens
  (For recommendations see the *How To Buy a Microscope* section after *Materials at a Glance* in the front of this book.)
- plastic microscope slides
- eye dropper
- pond water or protozoa kit

Protists (protozoa) can also be observed in hay water. To make hay water, cover a clump of dry hay with water and let it stand for several days at room temperature. Add water as needed.

As of this writing, the following materials are available from Home Science Tools, www.hometrainingtools.com:

- plastic microscope slides, MS-SLIDSPL or MS-SLPL144
- Basic Protozoa Set, LD-PROBASC

Objectives

In this unit students will look at pond water, hay water, or a protozoa kit to observe how protists (protozoa) move.

The objectives of this lesson are for students to:

- Make careful observations of protists moving.
- Practice using a microscope.

A microscope that is small and easy for young children to handle is recommended for this experiment. You may need to help your students learn how to look through a microscope lens. For practice, it might help to have the students look at larger objects, such as a piece of paper with lettering they can see. This will help the students orient their eyes for observing small things through the eyepiece. Before beginning the experiment, let them play with the microscope until they are comfortable using it.

Experiment

I. Think About It

Read this section of the *Laboratory Notebook* with your students.

The students have read about how protists move. Now have them think about what movement for a protist might look like and what looking at pond water through a microscope might show. Help them explore their ideas with questions such as:

- *What do you think pond water looks like?*
- *Will you see moving creatures?*
- *Do you think you will be able to tell if they are moving? How?*
- *Do you think you will see them rolling or twisting?*
- *Do you think they will swim fast or slow? Straight or in a circle?*
- *What else do you think you might observe in pond water?*

Have them draw what they think they will see when they look at pond water through a microscope. There are no right answers—just let students explore their ideas.

II. Observe It

Read this section of the *Laboratory Notebook* with your students.

This is mainly an observational experiment.

❶ **a)** Help the students set up the microscope. Placing the microscope on a flat, firm surface will make it easier to use.

b) Help the students put a drop of protozoa water (or pond water or hay water) on a plastic slide.

c) Help the students carefully place the slide in the microscope.

d) Help the students look through the eyepiece at the water on the slide.

It may take several tries before protists can be observed. Help students repeat setting up the slide with samples as many times as necessary.

It is important for students to practice observing as many different details as possible. Have them draw their observations.

❷-❺ There are several drawing frames in the *Laboratory Notebook* for students to fill in with drawings of the different features they observe in the pond water. Encourage them to spend plenty of time looking at all the different features they observe. You can encourage them to stay at the microscope by engaging them with questions such as:

- *What kind of protist do you think you are seeing?*
- *Is it moving fast or slowly? Can you see it spin?*
- *How does it stop? Can it move backwards?*
- *Do you see an amoeba?*
- *How fast does an amoeba move?*

❻-❼ Have students compare some of the protists they are observing. They are asked to make comparisons between different protists of the same kind (two paramecia, for example) and protists of different kinds (possibly a paramecium and an amoeba).

III. What Did You Discover?

Read this section of the *Laboratory Notebook* with your students.

Have the students answer the questions about the protists they observed. Encourage them to refer to their notes in the *Observe It* section and summarize their answers based on their observations. They should have been able to see different protists moving in different ways. Have them explain what their favorite protist was and why. Help them notice any differences between what they thought they would observe and what they actually observed.

IV. Why?

Read this section of the *Laboratory Notebook* with your students.

There are many different kinds of protists. Depending on what your students used for protozoa water, they should have been able to observe at least two different kinds of protists.

Protists move like sophisticated little machines. They roll and spin, stop and start, move forward, and back up. Explain to the students how remarkable protists are since they are made with only one cell yet can do so many different things.

V. Just For Fun

Have the students put some saliva on a slide and look at it under the microscope to see if they can find any organisms. Have them record their results.

Experiment 8

Little Creatures Eat

Materials Needed

- microscope with a 10X or 20x objective lens (For recommendations see the *How To Buy a Microscope* section after *Materials at a Glance* in the front of this book.)
- plastic microscope slides
- eye dropper
- pond water or protozoa kit
- small piece of chocolate

Optional

- baker's yeast
- Eosin Y stain
- distilled water

Protists (protozoa) can also be observed in hay water. To make hay water, cover a clump of dry hay with water, and let it stand for several days at room temperature. Add water as needed.

As of this writing, the following materials are available from Home Science Tools, www.hometrainingtools.com:

- plastic microscope slides, MS-SLIDSPL or MS-SLPL144
- Basic Protozoa Set, LD-PROBASC
- Eosin Y stain, CH-EOSIN

Objectives

In this unit students will look at pond water, hay water, or water from a protozoa kit to observe how protists (protozoa) eat.

The objectives of this lesson are for students to:

- Make careful observations of protists eating.
- Practice using a microscope.

Experiment

In this experiment students will focus on protists that are eating. If pond water or hay water is being used, there should be plenty of food for the protists to eat.

Baker's yeast stained with Eosin Y can be added to any of the kinds of protozoa water. The Eosin Y stained yeast will be ingested by the protists. Once ingested, the red stained yeast will turn blue. It may take some time for this observation.

To make baker's yeast and Eosin Y stain:

- Add 5 milliliters (one teaspoon) of dried yeast to 120 milliliters (1/2 cup) of distilled water. Allow it to dissolve. Let the mixture sit for a few minutes, then add one dropper of Eosin Y to one dropper of the baker's yeast solution and let sit for a few minutes.
- Look at a droplet of the mixture under the microscope. You should be able to see individual yeast cells that are stained red.

I. Think About It

Read this section of the *Laboratory Notebook* with your students.

The students have read about how protists eat. Have them first think about what it might look like for a protist to eat. Help them explore their ideas with questions such as:

- *How do you think a paramecium eats?*
- *Do you think you can watch it eat?*
- *Do you think you can tell if the food is going inside?*
- *How do you think an amoeba eats?*
- *Do you think an amoeba can eat fast moving food? Why or why not?*
- *What else do you think you might see as the protists eat?*

Have the students draw what they think they will observe through the microscope as they watch protists eat.

II. Observe It

Read this section of the *Laboratory Notebook* with your students.

❶ **a)** Help the students place a small droplet of the protozoa solution onto a microscope slide.

b) If using Eosin Y stained baker's yeast, have the students add a droplet of the stained baker's yeast to the protozoa water on the slide.

c) Help the students carefully place the slide in the microscope.

d) Have the students look through the eyepiece at the protozoa water on the slide.

(You may also position the slide correctly in the microscope and then add the liquids to it.)

It is important for students to practice observing as many different details as possible. Have them draw what they observe.

❷-❺ There are several drawing frames in the *Laboratory Workbook* for students to fill in with drawings of their observations of protists eating. Encourage the students to spend plenty of time looking at all the different features they observe. You can encourage them to stay at the microscope by engaging them with questions such as:

- *What kind of protist do you think you are seeing?*
- *Is it eating?*
- *Can you tell what it is eating?*
- *Can you tell if the protist is eating another protist or something else?*
- *How fast does it eat?*

❻-❼ Have the students compare some of the protists they are observing. They can make comparisons between different protists of the same kind (two paramecia, for example) and protists of different kinds (possibly a paramecium and an amoeba).

III. What Did You Discover?

Read this section of the *Laboratory Notebook* with your students.

Have students answer the questions about the protists they observed. Encourage them to summarize their answers based on their observations. They should have been able to see different protists eating. Have them explain what their favorite protist was, how it was eating, and why it was their favorite. Help them notice any differences between what they thought they would observe and what they actually observed.

IV. Why?

Read this section of the *Laboratory Notebook* with your students.

Different protists eat in different ways. Your students may or may not have been able to observe the protists eating. Explain to them that watching protists eat is sometimes like watching the tigers eat at the zoo. They may not be hungry. Repeat the experiment at a different time if your student was not able to observe protists eating.

V. Just For Fun

Help the students put a tiny piece of chocolate on the slide with the protozoa water. Have them look through the microscope to see if the protozoa will eat chocolate. Have them record their observations in the space provided.

Experiment 9

Oldy Moldy

Materials Needed

- 6-8 sealable plastic bags
- 2 pairs waterproof disposable gloves
- 2 or more pieces of newspaper or plastic
- 2 pieces of fruit
- 2-3 pieces of bread (will mold more quickly if it does not have preservatives)
- marking pen
- water

Optional

- colored pencils

Objectives

In this experiment, students will explore whether a mold that grows on fruit will grow on bread.

The objectives of this lesson are for students to:

- Describe events that happen during an experiment.
- Develop explanations using observations.

Experiment

I. Think About It

Read this section of the *Laboratory Notebook* with your students.

Have the students think about molds, mushrooms, and yeast they may have observed. Explore open inquiry with questions such as the following:

- *Have you ever seen a mushroom? What did it look like?*
- *Have you seen mushrooms that look different from each other? What did they look like?*
- *Have you ever seen mold? Where was it? What did it look like?*
- *Have you ever observed mold on fruit? Or mold on food that is in the refrigerator? Did it all look the same?*
- *Have you ever eaten bleu cheese? What did it taste like? What did it look like? Why do you think it is called bleu?*

Have the students answer the questions in this section of the *Laboratory Notebook*. There are no right answers to these questions.

II. Observe It

Read this section of the *Laboratory Notebook* with your students.

1. Molds require moisture and warm room temperatures to grow quickly and thrive. Help the students find a warm spot in the kitchen for their experiment and have them put down some newspaper or plastic for the fruit to sit on.

   Have the students cut the piece of fruit in half or do it for them. This will reduce handling of the moldy fruit later in the experiment. Have them let the fruit sit for several days until it is getting moldy. The fruit can be covered with a piece of newspaper or plastic to prevent the mold spores from scattering to other areas..

❷ Once the fruit is moldy, have the students take 3 plastic bags and label them **Control**, **Moldy Fruit**, and **Moldy Fruit + Bread** along with the date.

❸ Have the students put 5 milliliters (1 teaspoon) of water in each of the bags. If this amount of water does not make the food moist, have them add more water at this time. The goal is for the food to be moist but not soggy and for the bag to remain sealed during the experiment. As with making bread, this experiment might take longer in a cooler environment than a warmer one.

❹-❽ Have the students put on waterproof disposable gloves to handle the moldy fruit. They will put a fresh piece of bread in the **Control** bag, one piece of moldy fruit in the **Moldy Fruit** bag, and the other piece of moldy fruit and a fresh piece of bread in the **Moldy Fruit + Bread** bag. Have them thoroughly seal each bag and leave the bags in a warm, out of the way area.

Fungi do not need much fresh air to grow because they do not photosynthesize. The air in the bag is sufficient for a few days of cellular respiration of molds, which are small and grow slowly. Sealing the bags keeps the mold spores from being blown away and spreading around the kitchen.

The molds that grow on foods are generally not found to be dangerous to inhale. But to be safe, have the students cover the work area and clean up thoroughly. **Do not let them open the bags at the end of the experiment**—just throw them away.

❾ Have the students check the bags daily for 3-7 days. In the chart provided, have them write or draw each day's observations of the contents of each bag. **Do not let them open the bags.**

❿ When the experiment is over have the students **throw everything away without opening the bags.**

The molds the students observe may be a combination of the molds introduced during the experiment and molds already present on the food. Some fruit molds can grow on other types of foods and some cannot. Some fruits have anti-fungal agents, either naturally or added to them; molds will not grow on these.

The colors can vary with the types of molds that appear.

III. What Did You Discover?

Read this section of the *Laboratory Notebook* with your students.

Have the students answer the questions. These can be answered orally or in writing. Again, there are no right answers, and their answers will depend on what they actually observed.

IV. Why?

Read this section of the *Laboratory Notebook* with your students.

Discuss any questions that might come up.

V. Just For Fun

Here students will reverse the experiment to determine if bread mold will grow on fruit. To do this, they will first grow the mold on the bread instead of the fruit. A clean piece of fruit will be the control.

This time students can create their own experiment. To help them come up with the steps needed, have them review the steps of the experiment they performed previously. Guide them to come up with the experiment summarized below.

To have the bread get moldy instead of drying out, have the students put it in a plastic bag with 5 ml (1 teaspoon) of water. To reduce handling of the moldy bread, students can either cut in half the piece of bread that is to get moldy or let 2 pieces of bread get moldy.

Once the bread is moldy, students can cut in half the fresh fruit to be tested. Then they will label the bags and put the moldy bread and fresh fruit in them. Have them wear disposable waterproof gloves to handle the moldy bread. They will have a bag labeled **Control** that will contain a piece of the fresh fruit, a bag labeled **Moldy Bread** will contain a piece of the moldy bread, and a bag labeled **Moldy Bread + Fruit** will contain a piece of the moldy bread and a piece of the fresh fruit. The bags should be sealed thoroughly.

Have them check the bags each day **without opening them** and write or draw their observations in the chart provided.

At the end of the experiment, have students thoroughly clean the work area and **throw away the bags without opening them.**

Experiment 10

Measuring Time

Materials Needed

- Clock or stopwatch

Objectives

In this experiment, students will explore how to use a basic tool to measure an important physics parameter— time.

The objectives of this lesson are for students to:

- Use suitable tools, techniques, and quantitative measurements when appropriate.
- Use a simple tool to make measurements.

Experiment

I. Think About It

Read this section of the *Laboratory Notebook* with your students.

Have the students think about different events they could measure. Use questions such as the following to explore open inquiry and suggest possible events to measure.:

- *How long does it take you to go to school, to the grocery store, or to a park or museum?*
- *How long does it take you to brush your teeth?*
- *How long does it take you to eat breakfast?*
- *How long does it take you to walk around your house?*
- *How long does it take you to ride your bike down the block?*
- *How long does it take your mom or dad to walk the length of your backyard?*

Guide the students in coming up with 3 events they would like to measure and have them write these down. The instructions require that they time each event twice, so help them pick events they can measure more than once.

II. Observe It

Read this section of the *Laboratory Notebook* with your students.

Help the students use the clock or stopwatch. Have them record the time for an event and then repeat the event and record the time again. Make sure they record both the start time and end time. Then help them calculate the length of time for each occurrence.

Have them record any other observations they think are important or interesting. A box is provided for each event.

III. What Did You Discover?

Read this section of the *Laboratory Notebook* with your students.

❶-❻ Have the students answer the questions. These can be answered orally or in writing. There are no right answers and their answers will depend on what they actually observed.

IV. Why?

Read this section of the *Laboratory Notebook* with your students.

Discuss any questions that might come up.

V. Just For Fun

Have the students mark a distance they can run. Then have them measure how fast they can run that given distance. Have them repeat several times, recording the time it takes for each round and any other observations they find important or interesting. They can use a timer themselves or have someone else do it for them.

Experiment 11

Rolling Marbles

Materials Needed

- 1 small glass marble
- 1 large glass marble

Objectives

In this experiment students will explore inertia, mass, and friction.

The objectives of this lesson are for students to:

- Observe how mass and inertia are related and how the force of friction slows and eventually stops kinetic (moving) energy.
- Learn how to collect data and create a table of their results.

Experiment

I. Observe It

Read this section of the *Laboratory Notebook* with your students.

In this experiment students will perform a simple experiment to explore inertia, mass, and friction. The students will use two types of marbles — a small glass marble and a large glass marble (a marble that is several times larger than the small marble).

❶ Have the students take the small glass marble and the large glass marble and roll them on a smooth surface.

❷ Have the students observe how each marble rolls. Guide their inquiry with the following questions:

- *Do the marbles go straight?*
- *How far do the marbles go?*
- *Where do the marbles stop?*
- *How do the marbles stop?*

❸ Have the students record their observations. An example follows.

Small and Large Marbles Rolling on Smooth Surface

(Answers may vary.)

The small marble rolled across the wood floor from the chair to the couch. It stopped when it hit the couch, and it bounced back and stopped a few inches from the couch.

Have the students draw what they observed.

❹ Have the students take the same marbles and roll them across a carpeted surface or a grass lawn. This "rough" surface has more friction than a smooth surface.

❺ Again help the students think about their observations. Guide their inquiry with the following questions:

- *Do the marbles roll straight?*
- *How far do the marbles go?*
- *Does one marble go farther than the other?*
- *How do the marbles stop?*
- *Is one marble harder to roll than the other?*

❻ In the space provided, have the students record their observations. Have them pay particular attention to whether or not rolling the marbles on a rough surface is different from rolling them on a smooth surface. An example follows.

Small and Large Marbles Rolling on Rough Surface

(Answers may vary.)

The marble rolled across the carpeted floor from the desk to the chair. It did not roll as far as before. The marble just stopped on the carpet and did not hit anything. It was harder to roll the marble.

Have the students draw what they observed.

II. Think About It

Read this section of the *Laboratory Notebook* with your students.

❶ Review with the students Chapter 11 of the *Student Textbook* which covers mass, inertia, and friction. Help the students understand that friction is a force that stops moving objects. Have them apply this idea to the experiment they just performed to show which surface has more friction. Also help the students connect the concepts of mass and inertia — the more mass, the more inertia.

Help the students think about their experiment and any observations they made about the two marbles rolling. You can use the following questions to guide the discussion:

- *Based on your observations, was one marble easier to roll on the smooth surface?*
- *Based on your observations, was one marble easier to roll on the rough surface?*
- *Did one marble go farther than the other marble on the smooth surface?*
- *Did one marble go farther than the other marble on the rough surface?*
- *Once the marbles were moving, did one marble move faster or slower than the other marble?*

❷ Help the students decide which surface has more friction. If the students are not sure, review their results with them. Show them that the marbles had much more difficulty rolling on the rough surface than on the smooth surface and that this tells the students that the rough surface has more friction. They will place a circle around **Rough Surface.**

❸ Help the students decide which marble has the most mass. If the students are unsure, have them hold one marble in each hand. Can they feel which marble is heavier? The heavier marble has the most mass. The larger marble will be heavier than the smaller marble, so it has more mass. They will place a box around **Large Marble**.

❹ Help the students decide which marble has the most inertia. The marble with the most mass has the most inertia. Since the large marble has the most mass, it also has the most inertia. They will place a triangle around **Large Marble**.

III. What Did You Discover?

Read this section of the *Laboratory Notebook* with your students.

The questions can be answered verbally or in writing. With these questions help the students think about their observations. There are no "right" answers to these questions, and it is important for the students to write or discuss what they actually observed. Help them explore how the answer they got may be different from what they thought might happen.

Help the students compare the two marbles, and help them explore any similarities or differences between how the two marbles rolled on the different surfaces. They should have discovered that the larger marble is slightly harder to start rolling than the smaller marble, but that once it is going, it is harder to stop (it rolls farther, or takes more force to stop it from rolling). In their discussions, help them use the words "friction," "inertia," and "mass." For example, they might say that the small marble has less mass and less inertia than the large marble.

IV. Why?

Read this section of the *Laboratory Notebook* with your students.

The students compared two marbles of different sizes and how they roll on two different surfaces. They may have observed the larger marble rolling farther on the rough surface than the smaller marble. They may also have observed that the larger marble can roll farther than the small marble on the smooth surface. However, it may have taken a bigger push to get the larger marble to roll. Explain to them that what they observed is related to mass, inertia, and friction. The larger marble has more mass and more inertia, and so in order to stop, it requires more friction than the smaller marble requires. Also explain to the students that because of its inertia, a marble in outer space could keep moving without ever stopping because there is no air friction in outer space to stop it.

V. Just For Fun

Have the students observe what happens when a small marble hits a large marble and when a large marble hits a small marble. Ask them to observe if there is any difference, and if so, why. Have the students record their observations.

Experiment 12

Speed It Up!

Materials Needed

- stopwatch or clock
- an area to run in
- items for marking the beginning and ending of the running distance

Objectives

In this experiment, students will explore how to calculate speed using basic tools.

The objectives of this lesson are for students to:

- Use suitable tools, techniques, and quantitative measurements when appropriate.
- Use a simple tool to make measurements.

Experiment

I. Think About It

Read this section of the *Laboratory Notebook* with your students.

Have the students think about different distances they might run. If running is not possible for the students, explore other ways they can measure distance traveled over time. They could measure a rolling ball or how fast a pet can run a certain distance.

Explore open inquiry with questions such as the following:

- *Do you think you can run the length of the yard?*
- *Do you think you can run the length of the block?*
- *Do you think you can run the length of the football stadium?*
- *Do you think you can run the length of the city?*

Help the students think about distance, how far they might be able to run, and that shorter distances will be easier to run than longer distances. Help them pick a distance to run that is suitable for them. Again, if running is not possible, then have them select a different object they can measure, such as a rolling ball, a baseball being thrown, a bowling ball, how fast a pet runs, or how fast a parent, friend, or teacher can run.

II. Observe It

Read this section of the *Laboratory Notebook* with your students.

Help the students mark a distance they can run. Have them use their feet as the measuring tool by walking heel-to-toe and counting the steps. Explain that this won't be an accurate measurement but rather an estimation. "Experiencing" the length of a distance by using their own feet gives them a sense of space that using a ruler or measuring tape would not.

Have the student run the distance between the two points they mark. They can either hold the stopwatch or timer themselves, or you can time them. However, pick one method and stick to it for all five runs. Switching the way the time is measured can introduce error.

Help them record their results in the table provided.

Help the students calculate their average speed by adding all the speeds together and dividing by the number of runs (5 if they ran 5 times).

III. What Did You Discover?

Read this section of the *Laboratory Notebook* with your students.

❶-❻ Have the students answer the questions. These can be answered orally or in writing. There are no right answers and their answers will depend on what they actually observed.

IV. Why?

Read this section of the *Laboratory Notebook* with your students.

Discuss any questions that might come up.

V. Just For Fun

Read this section of the *Laboratory Notebook* with your students.

Have the students measure how fast you or a friend can run the same distance that was measured in the first part of the experiment for the same number of times. Have them use the timer or stopwatch to record the times in the chart provided. Then have them calculate the average speed and compare it to their own average.

Experiment 13

Keep the Train on Its Tracks!

Materials Needed

- 4 plastic or Styrofoam cups with the mouth larger than the base
- 2 long poles (dowels work well or any two long sticks that are the same thickness from end to end)
- tape
- a cylinder, 10-13 cm long (4-5 inches) [any cylindrical object, such as a pencil, a dowel, a cylindrical block, or a cylindrical drinking glass that is not tapered; a paper towel tube may be used if it is filled with sand and the ends taped]
- chalk

Objectives

In this experiment, students will explore how differential rotational motion keeps a train's wheels on the tracks.

The objectives of this lesson are for students to:

- Describe the motion of an object by tracing its position over time.
- Observe how a change in diameter (size) changes the rotational motion of an object.

Experiment

I. Think About It

Read this section of the *Laboratory Notebook* with your students.

If you live near a train station it would be useful to look at the train wheels of a physical train. If not, do a library or internet search to study the physical shape of a train wheel.

Explore open inquiry with questions such as the following:

- *How is a train wheel shaped?*
- *How is a train wheel different from a bicycle or car wheel?*
- *How fast do trains travel?*
- *How do you think a train stays on the tracks when it travels fast?*

Have the students answer the questions in this section. There are no right answers.

II. Observe It

Read this section of the *Laboratory Notebook* with your students.

If possible, use a cement surface or solid, flat floor for doing this experiment.

❶-❷ Have the students roll first an un-tapered cylinder and then a tapered plastic or Styrofoam cup on a flat surface. Each time, have them trace the path of the object with a piece of chalk. Then, in the spaces provided, have them record their observations of how and how far the object rolls.

❸ Help the students tape the 2 poles to the ground parallel to each other and about 5 cm (2") apart. Make sure the track made by the poles is narrower than the length of the cylinder and the taped plastic cups.

4. Have the students roll the cylinder on the poles. Have them observe how it moves and whether or not it stays on the poles or falls off. Have them record their observations in the space provided.

5. Have the students take 2 plastic or Styrofoam cups, put the open ends together and tape them end-to-end.

6. Have the students roll the taped plastic cups on the poles. Have them observe how they move and whether or not they stay on the poles or fall off. Have them record their observations in the space provided.

III. What Did You Discover?

Read this section of the *Laboratory Notebook* with your students.

1-8 Have the students answer the questions. These can be answered orally or in writing. Again, there are no right answers and their answers will depend on what they actually observed.

IV. Why?

Read this section of the *Laboratory Notebook* with your students.

Discuss any questions that might come up.

V. Just For Fun

Read this section of the *Laboratory Notebook* with your students.

1. Have the students think about what would happen if they reversed the position of the cups and taped the bottoms (smaller ends) together, then rolled them on the poles. Have them record their ideas in the space provided. There are no right answers.

2. Now the students will tape together the bottoms of two plastic cups and repeat the experiment. Have them record their results based on their observations.

Experiment 14

Smashing Hammers

Materials Needed

- plastic hammer
- regular metal hammer
- 3 pieces of banana
- 3 hardboiled eggs in the shell
- 3 raw potato halves
- 3 rocks of the same type and size (students can collect these)
- safety glasses

Optional

- 8 pieces of paper
- marking pen

Objectives

In this experiment students will explore how using different tools causes different outcomes.

The objectives of this lesson are to have students:

- Explore how different objects have different properties.
- Observe how the properties of different materials require the use of different tools.

Experiment

Before beginning the experiment have the students collect three rocks of the same type and similar size. The experimental results will vary depending on the hardness of the rocks and whether or not they are layered.

I. Think About It

Read this section of the *Laboratory Notebook* with your students.

Have the students think about the differences between a plastic hammer and a metal hammer. Also have them think about the differences between a piece of banana, a hardboiled egg in its shell, a raw potato, and a rock. They may answer the questions orally or in writing.

Guide open inquiry with questions such as:

- *Is a metal hammer heavier or lighter than a plastic hammer? Why?*
- *Do you think a hardboiled egg is harder or softer than a banana? Why or why not?*
- *Is a rock harder or softer than a hardboiled egg? Why?*
- *Do you think a plastic hammer can crush a rock? Why or why not?*
- *Do you think a metal hammer can crush a hardboiled egg? Why or why not?*
- *Do you think a metal hammer would work better for smashing a potato than a plastic hammer would? Why or why not?*

II. Observe It

Read this section of the *Laboratory Notebook* with your students.

In this experiment students will observe how using different tools can result in different outcomes. They will first smash a piece of banana with a plastic hammer and then smash another piece of banana with a metal hammer. This process will be repeated with hardboiled eggs in the shell, potato halves, and rocks. Students can write or draw their observations in the boxes provided, or they can relate their observations orally.

Have the students wear safety glasses to protect their eyes from possible debris.

Optional: Students can place each object on a piece of paper before smashing it and write or draw which hammer they'll be using.

Guide open inquiry with questions such as:

- *What happens to the piece of banana when you smash it with the plastic hammer? What does it look like after it's smashed? Why?*
- *What happens to the piece of banana when you smash it with the metal hammer? What does it look like? Does the hammer you use make a difference? Why?*
- *Does it make a difference whether you smash a hardboiled egg with the plastic hammer or the metal hammer? If so, what is different?*
- *Can you smash the potato with the plastic hammer? With the metal hammer? Is there a difference? Why or why not?*
- *What happens when you smash the rock with the plastic hammer? With the metal hammer? Does one of the hammers work better? Why or why not?*

III. What Did You Discover?

Read this section of the *Laboratory Notebook* with your students.

Have the students answer the questions in this section orally or in writing. Have them refer to their notes or drawings in the *Observe It* section. There are no right or wrong answers to these questions.

IV. Why?

Read this section of the *Laboratory Notebook* with your students.

Discuss any questions that might come up.

V. Just For Fun

Read this section of the *Laboratory Notebook* with your students.

Have the students use a magnifying glass or hand lens to look at each of the objects that is unsmashed, that was smashed with the plastic hammer, and that was smashed with the metal hammer. Students may write, draw, or state their observations orally.

Guide open inquiry with questions such as:

- *Did [the object] smash the way you thought it would? Why or why not?*
- *Does [the object] look different when you look at it through the magnifying glass (or hand lens) than it does when you look at it using only your eyes? Why or why not?*
- *Does the unsmashed [object] look different from the way the smashed one looks? How?*
- *Can you see a difference between the way [the object] looks when it was smashed with a plastic hammer and when it was smashed with a metal hammer? Why or why not?*

Experiment 15

All the Parts

Materials Needed

- a toy, small music box, or toy car that can be taken apart
- a second similar item that can be taken apart
- screwdriver
- small hammer
- other tools as needed

Note: The objects used in this experiment may not work again.

Objectives

In this experiment, students will explore how taking apart an object helps them learn in detail how the object works.

The objectives of this lesson are for students to:

- Use suitable tools to study an object.
- Observe how tools help scientists make better observations.

Experiment

Before starting the experiment

This experiment requires that the students disassemble an object to learn more about the individual parts. For this experiment choose a toy or small mechanical object that is OK to disassemble, understanding that the object may not work again when reassembled. Make sure the item is one that has a number of pieces that can be taken apart.

Inspect the object to be disassembled to see what tools will need to be provided to the students.

I. Think About It

Read this section of the *Laboratory Notebook* with the students.

Help the students think about the parts of a bicycle, a car, and an airplane. Discuss which parts are similar (e.g., wheels) and which parts are different (e.g., wings, motor, gears).

Explore open inquiry with questions such as the following:

- *What parts can you think of that are on a bicycle?*
- *What parts can you think of that are on a car?*
- *What parts can you think of that are on an airplane?*
- *Which parts do you think are similar and which do you think are different?*
- *Are there any parts that you think you would you find on all three objects? What are they?*
- *Are there any parts that you think you would find only on the bicycle? Only on the car? Only on the airplane?*

II. Observe It

Read this section of the *Laboratory Notebook* with your students.

❶ Have the students carefully observe the object. Guide them to note what the object does and how it functions.

In the box provided, have the students fill in the name of the object and write and/or draw their observations.

❷ Have the students carefully take apart the object, and guide them in making observations during disassembly. Have the students do as much of the disassembly themselves as possible. Once the object is taken apart, have the students count the number of parts and write it on the line provided.

❸ Students are to examine each part. Boxes are provided for them to draw the parts and to record the size, weight, and apparent function of each.

❹ Have the students reassemble the object. The object may or may not work when they are finished. Have them record their observations.

III. What Did You Discover?

Read this section of the *Laboratory Notebook* with your students.

Have the students answer the questions. There are no right answers and their answers will depend on what they actually observed

IV. Why?

Read this section of the *Laboratory Notebook* with your students.

Discuss any questions that might come up.

V. Just For Fun

Have the students take apart another object. Guide them in making good observations, and have them record their observations in the boxes provided.

Experiment 16

Edible Earth Parfait

Materials Needed

- 2 clear, tall glasses (drinking or parfait glasses)
- spoon (1 or more)
- 3-6 student-chosen food items that can be used to build a parfait model of Earth's layers (such as: graham crackers, peanut brittle, cookies, hot fudge, Jell-O, pudding, ice cream, cream cheese, cherry, nut, jelly bean, etc.)
- student-chosen inedible items that can be used to build a parfait model of Earth's layers (such as: rocks, mud, dirt, clay, dog or cat food, Legos, etc.)
- colored pencils

Objectives

In this experiment, students will explore how models help scientists make educated guesses about how things work.

The objectives of this lesson are for students to:

- Use suitable tools to study an object.
- Observe how tools help scientists make better observations.

Experiment

I. Think About It

Read this section of the *Laboratory Notebook* with your students.

Help your students think about different foods they could use for the crust, lithosphere, asthenosphere, mesosphere, outer core, and inner core. Students may combine layers with similar qualities. Solid, thin food items like crackers, peanut brittle, or cookies might be good items for the crust. Softer food items like hot fudge, Jell-O, pudding, or ice cream might be good items for the inner layers. A cherry, nut, jelly bean, or similar item could be used for the inner core. Help your students think about the consistency of different foods and whether or not they might make a good representation of a particular layer of an edible Earth.

Explore open inquiry with the following:

- *What are some solid food items you like?*
- *What are some soft food items you like?*
- *What are some combinations of food items you like?*
- *How well do you think different food items will fit together and taste?*
- *Which layers of Earth do you want to include in your parfait? All of them? If not all, which ones? Why?*
- *Which layers would you represent with solid (hard) food? Which would you make with softer foods? Why?*

II. Observe It

Read this section of the *Laboratory Notebook* with your students.

❶ Help the students plan a layered Edible Earth Parfait. This experiment requires that the students use a variety of food items. Guide students to pick foods that work for you and for them. Have them list the foods and which layer each food represents.

❷ Have students think about whether they were able to find foods to represent the properties of the different the layers of Earth.

❸-❹ Have the students assemble their Edible Earth Parfait and make careful observations about it. Help them observe whether the layers are interacting.

❺ The students can now eat their model of Earth's layers.

❻ Have the students look up the definition of parfait in a dictionary or online. Ask them if they think their edible Earth model fits the definition of a parfait. Why or why not?

III. What Did You Discover?

Read this section of the *Laboratory Notebook* with your students.

Have the students answer the questions. There are no right answers, and their answers will depend on what they actually observed.

IV. Why?

Read this section of the *Laboratory Notebook* with your students.

Discuss any questions that might come up.

V. Just For Fun

Have the students review *I. Think About It* and choose inedible items to create another model of Earth's layers. A box is provided for them to draw and label their model. Have them observe any similarities or differences between the edible and inedible models. Are the layers interacting in this model?

Experiment 17

What's the Weather?

Materials Needed

- colored pencils
- outdoor thermometer
- helium-filled balloon
- string

This experiment is done over the course of a week.

Objectives

In this weeklong experiment students will explore the atmosphere by making daily observations of the weather and its effects.

The objectives of this lesson are for students to:

- Observe daily changes in the weather.
- Observe how changes in the weather can affect plants, animals, and the land.

Experiment

I. Think About It

Read this section of the *Laboratory Notebook* with your students.

Have the students think about what the weather is like where they live and how it changes from day to day and from season to season. Guide open inquiry using questions such as:

- *What do clouds look like? Do they always look the same? Why or why not?*
- *Do you think changes can happen to the ground when it rains hard? Why or wy not?*
- *What do you think happens to plants when there is a breeze? When the wind blows really hard? Why?*
- *Do you think wind can make changes to the land? Why or why not?*
- *Do you think animals act differently when it is raining than when it is sunny? When it is very windy? When it is very hot or very cold? When it is day or night? Why or why not?*
- *Do you think you act differently in different kinds of weather? During the daytime or at night? In summer and winter? Why or why not?*

II. Observe It

Read this section of the *Laboratory Notebook* with your students.

Have the students observe the weather at about the same time every day for a week. Have them record the temperature, and write, draw, or relate orally their observations about the weather. Guide open inquiry with questions such as:

- *What do you see when you look at the sky? Is it the same color every day? Why or why not?*
- *What do you notice about clouds? What do they look like? Do they look the same every day? Why or why not?*
- *Is the temperature the same every day? Is it the same at night? Why or why not?*
- *What differences do you observe in animal behavior as the weather changes? When it's sunny? When it rains? When it's windy?*
- *Do changes in the weather affect the plants? When it's sunny? When it rains? When it's windy? Why or why not?*
- *Do you observe any changes to the ground when it rains hard or is very windy? Why or why not?*

III. What Did You Discover?

Read this section of the *Laboratory Notebook* with your students.

Have the students review their notes from the *Observe It* section and answer the questions based on their observations. There are no right answers.

IV. Why?

Read this section of the *Laboratory Notebook* with your students.

Discuss any questions that might come up.

V. Just For Fun

Read this section of the *Laboratory Notebook* with your students.

In this experiment students will make a simple tool to measure the wind.

Help the students find an object outdoors that they can tie the balloon to. The balloon should have enough space around it that it won't hit anything if the wind blows hard. Have the students observe the balloon several times during the day.

Guide open inquiry with questions such as:

- *Do you think you will be able to test the wind by using a balloon? Why or why not?*
- *What things do you think the balloon might be able to measure?*
- *By looking at the balloon, can you tell if the wind is blowing? Why or why not?*
- *Can you tell how hard the wind is blowing? Why or why not?*
- *Can you tell the direction of the wind? Why or why not?*
- *Do you think a balloon is a good way to measure the wind? Why or why not?*

Experiment 18

Building a Telescope

Materials Needed

- colored pencils
- clear night sky
- basketball or other large object(s)

Telescope materials*

- empty cardboard paper towel tube
- 1-2 sheets of card stock
  or
  1 manila file folder cut in half
- tape
- 2 lenses with different focal lengths
  Home Science Tools
  Item# OP-LEN4x15 and
  Item# OP-LEN4x50
  http://www.hometrainingtools.com

* Alternatively, you can look online for a telescope kit

Objectives

In this unit, students will build a simple telescope and explore how it can be used to make faraway objects appear larger with more detail.

The objectives of this lesson are to help students:

- Build a working telescope.
- Practice using a scientific tool and observe how using the tool changes what can be explored.

Experiment

I. Think About It

Read this section of the *Laboratory Notebook* with your students.

Gather all the materials needed for building the telescope.

Have the students take the sheet of card stock or 1/2 manila file folder and roll it into a tube with the longest side of the paper being the length of the tube. Have them tape it to hold it together. This, along with the cardboard paper towel tube, will make the barrel of the telescope.

Next, have the students examine the different pieces of the telescope.

Use the following questions to help them think about the various parts of the telescope.

- *Which items are the lenses?*
- *How many lenses are there? Why?*
- *Why are the lenses clear?*
- *Which items are the tubes?*
- *Why do you think there are different tubes?*
- *Where do you think the lenses will go?*

II. Observe It

Read this section of the *Laboratory Notebook* with your students.

Have the students assemble the telescope.

❶ Show them which lens is the eyepiece lens (shorter focal length) and have them tape this lens into one end of the card stock or file folder tube. They will need to either adjust the diameter of the tube they taped together previously, or if it seems easier, make a new tube. Have them tape the lens securely while being careful not to cover too much of the lens with tape.

❷ Have the students tape the objective lens, the longer focal length lens, to one end of the paper towel tube.

❸ Have them slide the open end of the heavy paper tube into the open end of the paper towel tube. This completes the telescope.

❹ Students are now to look at a faraway object, first using just their eyes and then by looking at it through the eyepiece of the telescope. The object they are viewing will appear upside down. Explain to the students that more complicated telescopes contain mirrors to turn the image right side up.

Have them experiment with sliding in and out the tube that has the objective lens. Help them observe that this will adjust the focus. If they are unable to focus the telescope, have them make one of the tubes longer by taping more card stock onto it.

❺ Students are to use the telescope to observe several objects that are far away. Have them pick objects that they can see while using only their eyes. Have them look at both small objects and large objects. Some suggestions:

- *a car parked across the street*
- *a building several blocks away*
- *a tree or flower at the far end of the yard or park*
- *a mountaintop or other geological feature*

Have the students compare the details they observe while using only their eyes and the details they observe when using the telescope. Have them note what kinds of details they can observe with the telescope that they cannot observe with their eyes only and draw their observations.

❻ Now that they have practiced with the telescope, have the students pick several stars, the Moon, or other features to observe in the night sky. Have them take their time observing through their telescope. Part of scientific discovery is slowing down and simply observing. Have them draw what they see.

III. What Did You Discover?

Read this section of the *Laboratory Notebook* with your students.

The questions can be answered verbally or in writing depending on the writing ability of the student. With these questions, help the students think about their observations. There are no "right" answers to these questions, and it is important for the students to write or discuss what they actually observed.

IV. Why?

Read this section of the *Laboratory Notebook* with your students.

Even when using scientific tools, the most important part of astronomy is observing. Good observing takes patience and practice. Ask the students to explain how using the telescope changes what they can observe.

Have the students imagine what it would be like to be Galileo. What would it have been like to be the first person to see planets in detail? How much more information about stars and planets can we explore today because of the telescope?

V. Just For Fun

A basketball is suggested for this experiment, but a different large object with a patterned or textured surface can be used. Help the students note as many features as possible about the object. Discuss with them any differences they notice when the object is held close, is far away and seen with only their eyes, and when it is far away and seen through the telescope. Ask which way they were able to get the most information. The least? Was there information they could not obtain when the object was far away?

The experiment may be repeated by moving the object different distances away or by using different objects.

Experiment 19

Tracking a Constellation

Materials Needed

- colored pencils
- night sky
- daytime sky or textured surface

Optional

- book or online information about constellations
- globe or basketball

Objectives

In this experiment students will observe a constellation of their choice to see whether its position changes in the night sky over the course of a week.

The objectives of this lesson are to have students:

- Make careful observations.
- Use their eyes as a tool in a scientific experiment.

Experiment

I. Think About It

Read this section of the *Laboratory Notebook* with your students.

Have the students answer the questions in this section. There are no right answers. Guide open inquiry with questions such as the following.

- *Do you think every star you can see is part of a constellation? Why or why not?*
- *Do you think there are lots and lots of constellations? Why or why not?*
- *Do you think people are coming up with new constellations all the time? Why or why not?*
- *If you look at the sky at different times of the night, do you think you will always see the stars in the same position? Why or why not?*
- *Do you think you would see the same groups of stars no matter where you are on Earth? Why or why not?*
- *From what location on Earth do you think you could see the most constellations? Why?*

II. Observe It

Read this section of the *Laboratory Notebook* with your students.

❶ Have the students pick a constellation to observe. To find more constellations than those mentioned in the *Student Textbook*, they can consult a book about constellations or look online for more information. Guide them in selecting a constellation that will be visible at the time they will be looking for it.

2. Help the students locate the constellation they have chosen to observe. If they can't find it on the first night, have them try again the next night or several nights until they can see it.

3. Have the students record the time and date when they first see their constellation. For the following six days they will view the constellation at the same time.

4. A box is provided for students to record the position of the constellation by drawing or writing. Have them observe how high in the sky the constellation is and in which direction. It can be helpful to have them note a landmark to use to track the relative position of the constellation during the experiment—for example, how the constellations is positioned over a fence post, tree branch, or corner of a building.

5. Have the students observe the constellation at the same time for six more days and record their observations about its location. If it's too cloudy to see the constellation, they can either note this for that day's observation or they can observe the constellation on six clear nights even if there are days in between the observations.

III. What Did You Discover?

Read this section of the *Laboratory Notebook* with your students.

Have the students answer the questions. Answers will be based on their observations.

IV. Why?

Read this section of the *Laboratory Notebook* with your students. Answer any questions that may come up.

If you have a globe, it can be used to demonstrate how the spin of Earth on its axis changes the view of the constellations during the course of a night. A globe can also be used show how the constellations that are visible at any one time varies according to one's location on Earth and how most or possibly all constellations will be visible from the equator. In addition, the globe can be used to show how Earth's orbit around the Sun changes Earth's position relative to the constellations and thus changes our view of the stars over the course of a year. A basketball or other ball may be used instead of a globe.

Although over time the stars do change their position in the universe relative to Earth, this happens so slowly that it isn't obvious in a lifetime.

V. Just For Fun

In this experiment students look at clouds and use their imagination to find shapes that remind them of some person, animal, or other object. Then they are asked to draw a picture of what they see and write a short story about it. The story can be one sentence or longer—wherever their imagination takes them. It can be fiction or nonfiction.

If there are no clouds, help students find a textured surface that provides enough variation to suggest different shapes.

Experiment 20

Modeling an Orbit

Materials Needed

- Styrofoam ball
- pick, awl, or other thin, sharp object to poke a hole through the center of the ball
- nylon string
- scissors
- 2 or more marbles of different sizes
- cups that are different sizes

Objectives

In this unit, students will observe how two opposing forces keep a Styrofoam ball in a circular orbit.

The objectives of this lesson are for students to:

- Create a model of a planetary orbit.
- Explore opposing forces.

Experiment

I. Think About It

Read this section of the *Laboratory Notebook* with your students.

Have the students think about what happens when a string is used to whirl a ball in the air. Use questions such as the following to guide their inquiry.

- *What do you think will happen when you hold the end of the string and whirl the ball in a circle? Will the ball stay in the same position on the string?*
- *Do you think the ball will move towards or away from your hand when you whirl it? Why?*
- *Do you think the ball will fly off the end of the string when you whirl it? Why or why not?*
  (This will happen if the knot isn't large enough.)
- *What will happen if you shorten the string? Will the ball move faster or slower with a shorter string?*

II. Observe It

Read this section of the *Laboratory Notebook* with your students.

❶ The students are to assemble the ball and string. Help them pierce the ball with a pick or sharp tool.

❷ Have the students tie a large knot at one end of the string and then thread the nylon string through the ball. When the ball and string are assembled, the ball should be near the unknotted end of the string with enough string at this end for the student to grasp firmly. The ball should be able to slide on the string, and the knot should be large enough that the ball won't come off the end of the string when it is whirled.

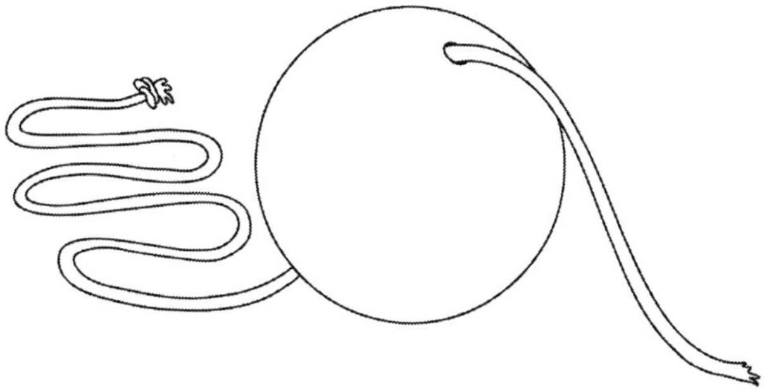

❸ It is useful to use the floor as a reference. Have the students whirl the ball around their hand with their hand fixed in the center of the circle. As they whirl it, the ball will slide outward along the string until it reaches the knotted end. The ball will then follow a circular path that stays at the same distance from their hand.

❹-❺ Have the students shorten and lengthen the string and observe how this changes the way the ball moves.

Encourage open inquiry with the following questions:

- *Does the ball go around faster or slower when the string is short?*
- *Is it easier or more difficult to spin the ball with a shorter string?*
- *If you slow down the speed at which you are spinning the ball, what happens to the ball?*
- *How fast can you spin the ball?*

III. What Did You Discover?

Read this section of the *Laboratory Notebook* with your students.

The questions can be answered verbally or in writing depending on the writing ability of the student. With these questions, help the students think about their observations. There are no "right" answers to these questions, and it is important for the students to write or discuss what they actually observed.

IV. Why?

Read this section of the *Laboratory Notebook* with your students.

There are two opposing forces that keep planets in a circular orbit around the Sun. One force, created by the speed and momentum of the planet, pushes the planet outward. The other force, the gravitational force of the Sun, pulls the planet inward. These two forces balance to keep the planets in circular orbits.

The balance of opposing forces is simulated in this experiment. As the ball is spun in a circle, a force causes it to travel outward. When it reaches the end of the string, the ball is pulled back towards the center, but since it is still rotating, it is also being pulled outward. The balance between the outward and inward forces keeps the ball in a circular orbit.

Planetary orbits are not quite circular, and the planets actually speed up as they near the Sun and slow down as they get farther away.

V. Just For Fun

Help the students use a marble in a cup to model an orbit. As the student moves the cup in a circular motion, the marble begins to circle the inside of the cup. In this experiment the cup creates an inward force on the marble while the outward force of the marble's momentum pushes against the cup. The two forces (inward and outward) are in balance which results in the marble circling the cup much as a planet orbits the Sun.

Have the students try using different size marbles, one at a time, in the same cup. Then they can try the marbles in different size cups. Help them notice any differences that may occur in the way the marbles move.

Experiment 21

Brightest or Closest?

Materials Needed

- flashlight with new batteries
- glow sticks in assorted colors may be found in places such as Walmart, toy stores, and online

Objectives

The brightness of a star depends on how much light energy the star generates and not necessarily how close the star is to Earth. In this unit, students will observe the luminosity of two different light sources—a flashlight and a glow stick—to demonstrate that the brightest shining object may not be the closest.

The objectives of this lesson are for students to:

- Observe how different light sources have different luminosity (light energy output).
- Explore what happens when two light sources of different luminosity are different distances away.

Experiment

I. Think About It

Read this section of the *Laboratory Notebook* with your students.

Before the students perform the experiment, have them think about the differences between a glow stick and a flashlight. Draw on any previous knowledge they have about using flashlights or glow sticks.

Use the following questions to help guide the students' inquiry:

- *Which one (glow stick or flashlight) do you think is brighter?*
- *What happens to a flashlight when the batteries run down?*
- *How long do you think a glow stick will stay lit?*
- *If you put the glow stick and flashlight side-by-side and observe them from far away, do you think you would see both?*
- *What do you think will happen if the glow stick is closer to you? Will it look brighter than the flashlight?*

II. Observe It

Read this section of the *Laboratory Notebook* with your students.

❶ Help the students bend their glow stick so that the inner chamber cracks. This does not take much force. Too much force can cause the plastic to break, spilling the contents. Perform the experiment immediately. The glow stick lights up due to a chemical reaction and will only produce light for a few hours.

❷- ❹ Have the students observe how far into the distance the glow stick and the flashlight illuminate. This works best in a dark room or on a moonless, dark night. It is helpful to have the students use a point of reference. For example, you can have them stretch out their free hand and observe if they can see all their fingers in the light from the glow stick and then from the flashlight. Then have them look beyond the end of their hand to some object a little farther away and repeat the experiment with that object and so on until the light from the glow stick or flashlight is no longer bright enough to illuminate an object.

❺ Have the students place the glow stick and flashlight side-by-side on the ground or floor. Have the students walk about a meter (several feet) to several meters (yards) away from both and look toward the glow stick and flashlight without looking directly at the flashlight. If the flashlight is strong enough, it will likely wash out all of the luminosity from the glow stick. This simulates what happens when bright stars wash out the appearance of dim stars.

❻-❼ Have the students place the flashlight a meter or so (several feet) behind the glow stick and repeat their observations. The flashlight, being brighter, will appear closer than the glow stick.

Encourage open inquiry with the following questions.

- *Is the flashlight or the glow stick brighter?*
- *Why do you think the flashlight creates more light energy than the glow stick?*
- *How long do you think the glow stick's light will last?*
- *How long do you think the flashlight's light will last?*

III. What Did You Discover?

Read this section of the *Laboratory Notebook* with your students.

The questions can be answered verbally or in writing depending on the writing ability of the student. With these questions, help the students think about their observations. There are no "right" answers to these questions, and it is important for the students to write or discuss what they actually observed.

IV. Why?

Read this section of the *Laboratory Notebook* with your students.

Stars that appear closer may actually be farther away but brighter. By observing two different light sources of different luminosity, students can begin to understand how bright stars that appear closer or larger may in fact be farther away than smaller, dimmer stars because of their luminosity or light energy output.

Glow sticks produce light through a chemical process called chemiluminescence. When the inner chamber of a glow stick is broken, two chemicals are allowed to mix and react with each other. When this happens, fluorescent light is emitted. The luminosity of a glow stick is several times lower than that of flashlight.

V. Just For Fun

Have the students perform the experiment using different colored glow sticks and compare the results to their original experiment. They can also compare the glow sticks to each other to try to determine if one color is brighter than another. Ask them if they think that by combining the light from several glow sticks they will get a light that is as bright as the flashlight. Have them try it.

Experiment 22

Teamwork

Materials Needed

- 10 small pieces of paper
- a box for the paper pieces
- 2 beakers or jars—one with 118 ml (1/2 cup) of vinegar and one with 118 ml (1/2 cup) of baking soda water (5 ml [1 tsp] baking soda in 118 ml (1/2 cup) water)
- magnifying glass (1-2)
- 2 balls of different weights (e.g., a glass marble and a metal marble, a plastic ball and a baseball)
- rock hammer or regular metal hammer
- safety glasses
- garden trowel or large metal spoon
- 10 pieces of paper
- 5 pens or pencils
- 4 friends or family members to help with the experiment
- scissors

Objectives

In this experiment, students will learn how to work together as a team and examine the world around them from multiple scientific perspectives.

The objectives of this lesson are to have students:

- Use different kinds of investigations
- Review and ask questions about the results of other scientists' work.

Experiment

Setup

In a beaker or jar, mix 5 ml [1 tsp] baking soda into 118 ml (1/2 cup) water. Measure 118 ml (1/2 cup) vinegar into a separate beaker or jar. Or you can have the experimenter who is doing the chemistry part of the experiment do this.

I. Think About It

Read this section of the *Laboratory Notebook* with your students.

Have the students think about how much scientific information exists today and how difficult it would be to learn all of it.

Students may have a difficult time understanding the volume of information available today. To help them visualize how much information is available, have them write down ten things they know. Give them ten small pieces of paper and have them write one thing on each. Then have the students place the ten pieces of paper in a box.

Explore the idea of large amounts of information with the following questions

- *How big would your box need to be to hold 100 pieces of paper?*
- *How big would your box need to be to hold 1,000 pieces of paper?*
- *How big would your box need to be to hold 10,000 pieces of paper?*
- *How big would your box need to be to hold a million pieces of paper?*
- *There are billions of scientific "facts" — how big would your box need to be to hold all of them?*
- *How easy do you think it would be to learn all those facts?*

Have them draw their idea of how big the box would have to be to hold all the known scientific facts. They can put another object in their drawing for size reference.

II. Observe It

Read this section of the *Laboratory Notebook* with your students.

1. Help your student select four friends or family members who can be part of their team of scientists. The student will select the core science subject they would like to experiment with and assign subjects to the other four participants. The subjects are chemistry, biology, physics, geology, and astronomy.

2. Have the student provide each participant with the tools they need.

   You will provide each scientist with written instructions for their part of the experiment. The instructions are provided at the end of this chapter. Cut them apart and give instructions to the appropriate team member. Make sure the participants don't share their instructions at this point.

   Give them each a piece of paper and a pen for recording their observations.

3. Have the participants imagine they are on a planet that is similar to Earth and is being visited for the first time. They are the scientists who will be performing the first tests. Have them try to imagine they are not still on Earth.

4. Help your student choose an area to explore, such as a backyard, park, or other open area. The chemist and geologist will need to be able to dig small samples of dirt. The physicist will need a flat, solid surface to drop balls on.

   Have your student direct the team members to explore the area while following the instructions you provided.

   Each team member will explore the area from the perspective of their specific science subject. Below are explanations of the experiments each participant will perform.

   Chemist

   The chemist will test the pH of the surrounding soil. The chemist will begin by scooping some soil and placing it in the beaker with the **vinegar**. If the soil bubbles or fizzes, the chemist can record that the "**soil is alkaline (basic)**" because it reacts with the vinegar. Then the chemist will scoop fresh soil and place it in the **baking soda/water** mixture. If the soil bubbles or fizzes, have the chemist record that the "**soil is acidic**" because it reacts with the baking soda. If nothing happens in either case, the chemist will record that the soil is "**neutral**."

   Biologist

   The biologist will use the magnifying glass to observe nearby plant and animal life and will record the details of what they see.

Physicist

The physicist will hold two balls a few feet above the ground and let them go at the same time to observe if the balls strike the ground at the same time. This experiment will be repeated three times, and observations will be recorded each time. The balls should strike the ground at the same time, and the physicist can conclude that gravity works the same way on the new planet as it does on Earth.

Geologist

The geologist is to break open some rocks with the hammer and/or dig into the ground with a hammer. The geologist can use a magnifying glass to observe the rocks and soil. Have them record their observations.

Astronomer

The astronomer will pretend that they are on the newly discovered planet. They will use their imagination to determine how many moons the new planet has, how many suns are nearby, if the new planet is spinning like Earth, and any other "observations" they think of.

❺-❻ When all the team members are done with their part of the experiment, have your student direct them to share what they learned. In the spaces provided, have your student record the participants' observations.

III. What Did You Discover?

Read this section of the *Laboratory Notebook* with your students.

Have the students answer the questions. There are no right answers and their answers will depend on what was actually observed.

IV. Why?

Read this section of the *Laboratory Notebook* with your students.

Discuss any questions that might come up.

V. Just For Fun

Read this section of the *Laboratory Notebook* with your students.

Have the experimenters trade roles and repeat the experiments.

Have your student record the results and compare them to the previous results to see if there are any differences. Have them discuss any differences that might have occurred.

Cut apart the instructions on the following page and provide them to the experimenters.

Chemist

Perform a test of the soil of this newly discovered planet.

Scoop some soil and place it in the beaker that contains vinegar. If the soil bubbles or fizzes, record that the "soil is alkaline (basic)."

Next scoop fresh soil and place it in the baking soda/water mixture. If the soil bubbles or fizzes, record that the "soil is acidic."

If nothing happens in either case, record that the soil is "neutral."

Biologist

Observe nearby plant and animal life on this newly discovered planet.

Use a magnifying glass to look closely at some plants and animals that you see.

Record your observations.

Physicist

Test the gravity on this newly discovered planet.

Hold two balls a few feet above the ground and let them go at the same time. Observe whether the balls strike the ground at the same time. Repeat the experiment three times, and record your observations each time.

Geologist

Observe the features of the rocks and soil of this newly discovered planet.

Put on safety glasses and use a hammer to break open some rocks and/or dig in the ground. Look closely at the rocks and soil. Record your observations.

Astronomer

Observe this newly discovered planet from an astronomical point of view. Record how many moons the new planet has, how many suns are nearby, and if the new planet is spinning like Earth. Can you observe anything else?

(Use your imagination to think about this newly discovered planet.)